3 North Carolina NC EOG Grade 3 Math Practice Tests

Full-Length Test Prep with Detailed Answer Explanations

Dr. A. Nazari

3 Practice Tests to Get You Started!

Hey there, future math whiz! 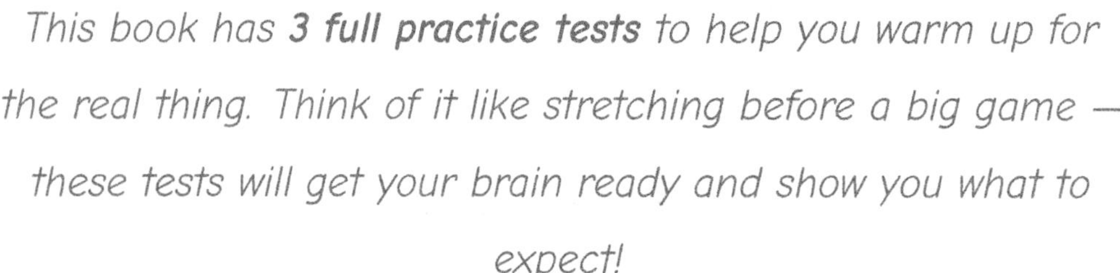

This book has **3 full practice tests** to help you warm up for the real thing. Think of it like stretching before a big game — these tests will get your brain ready and show you what to expect!

👍 Three tests is the **perfect start**!

👍 Each one helps you feel **more ready**!

👍 You'll be surprised how much you **already know**!

Sharpen your pencil and let's get warmed up! 🔥

> **❝** Three practice tests is a great way to start. Take your time with each one, and you'll feel more confident every step of the way! **❞**

📕 How to Use This Book 📕

What's Inside This Book

- **3 Full-Length Practice Tests** — Each one covers all the Grade 3 math topics you need to know!

- **Answer Key with Explanations** — Find out why each answer is correct, not just what the answer is.

- **Reference Pages** — A math symbols chart and multiplication table you can peek at any time.

- **A Test Tracker** — Write down your scores and watch your confidence grow!

A Simple 3-Test Plan

With just 3 tests, here's a great way to use them:

- **Test 1** — **The Warm-Up.** Take this test without a timer. Get comfortable with the question types. Don't worry about your score — just do your best!

- **Test 2** — **The Practice Round.** After reviewing Test 1, try this one with a timer (ask a grown-up!). Focus on the topics that were tricky last time.

- **Test 3** — **The Real Deal.** Treat this like the actual test: quiet room, timed, no peeking at answers. See how much you've improved!

Multiple Choice

Pick the **one best answer** from choices A, B, C, or D. Not sure? Cross out the ones you know are wrong, then pick from what's left. That's a smart move!

Short Answer

Write your answer **and** show your work! Even if your final answer isn't right, showing your steps can earn you credit. Use scratch paper if you need more room.

66 After Each Test 99

Flip to the Answer Key and check your work. For every question you got wrong, **read the explanation carefully.** Then write the tricky topics on your Test Tracker page. If you need extra help, grab our **Grade 3 Math Study Guide!**

> ⭐ **Fun fact:** Three tests is all it takes to see real improvement! Most kids feel way more confident after just a few rounds of practice. ⭐

Find more at
ViewMath.com/NC-Grade3

 Tips for Test Day

🌙 The Night Before

- ✅ **Sleep early** — your brain learns while you sleep!
- ✅ **Pack your supplies** — pencils, eraser, scratch paper, all ready to go.
- ✅ **Tell yourself:** "I've been practicing. I'm going to do great!"

👍 5 Simple Rules for Every Test

1. **Read the question twice.** The first time to understand it. The second time to catch details.
2. **Show your work.** Write the steps down, even on scratch paper. It helps you think!
3. **Skip the hard ones.** Put a small star next to tricky questions and come back later. Answer the easy ones first!
4. **Never leave a blank.** For multiple choice, your best guess is better than no answer at all.
5. **Check your work.** Finished early? Go back and re-read your answers.

✅ Smart Moves

- Take a deep breath before you begin
- Underline key words in the question
- Use drawings or number lines to help
- Cross out wrong answers first
- Double-check addition and subtraction

❌ Traps to Avoid

- Rushing and not reading carefully
- Picking the first answer that "looks right"
- Forgetting to carry or borrow numbers
- Skipping a question permanently
- Panicking when you see a tough problem

> 66 Remember, the very first practice test is the hardest — not because the questions are harder, but because everything is new! By Test 3, you'll feel like a pro. Trust me! 99

 Get Ready to Practice 

Pencils	**Eraser**	**Scratch Paper**
Sharpened and ready!	Everyone makes mistakes!	For working things out

A Calm Spot	**A Grown-Up**	**A Can-Do Attitude**
Somewhere quiet to focus	To help set a timer	You've totally got this!

✓ Allowed During Tests

- Pencils and erasers
- Blank scratch paper
- The **reference pages** in this book
- A ruler (for measurement questions)

✗ Not Allowed

- Calculators
- Phones, tablets, or computers
- Help from anyone else
- Your study guide (save it for after!)

👥 For Parents & Teachers

- With only 3 tests, **space them at least a week apart**. This gives time to review mistakes before trying the next one.
- Let your child take Test 1 untimed to build familiarity.
- After each test, go through the Answer Key together. Focus on **understanding the "why,"** not just the score.
- If a topic keeps tripping them up, review it in our **Grade 3 Math Study Guide** before the next practice test.
- Celebrate every bit of progress — even getting one more question right is a win!

X^1 Math Reference Sheet X^1

You may use this page during your practice tests!

Symbol	Name	What It Means	
$+$	Plus (Add)	Put numbers together.	$3 + 5 = 8$
$-$	Minus (Subtract)	Take away from a number.	$9 - 4 = 5$
$\times$	Times (Multiply)	Add equal groups.	$4 \times 3 = 12$
$\div$	Divide	Split into equal groups.	$12 \div 3 = 4$
$=$	Equals	Both sides are the same.	$2 + 3 = 5$
$>$	Greater Than	The left number is bigger.	$7 > 3$
$<$	Less Than	The left number is smaller.	$2 < 9$
$\frac{1}{2}$	Fraction Bar	Part of a whole.	$\frac{1}{2}$ means 1 out of 2 equal parts

📖 Key Math Words

- **Sum** — the answer when you add
- **Difference** — the answer when you subtract
- **Product** — the answer when you multiply
- **Quotient** — the answer when you divide
- **Factor** — a number you multiply
- **Array** — objects in rows and columns
- **Fraction** — a part of a whole
- **Numerator** — the top number in a fraction
- **Denominator** — the bottom number
- **Equation** — a math sentence with $=$
- **Estimate** — a smart guess, close to the real answer
- **Perimeter** — the distance around a shape
- **Area** — the space inside a shape
- **Rounding** — making a number simpler by going to the nearest ten or hundred

🔍 Word Problem Clue Words

- **Add** (+): in all, total, altogether, combined, sum, both, more
- **Subtract** (−): how many more, how many left, fewer, difference, remain
- **Multiply** (×): each, every, groups of, times, rows of, per
- **Divide** (÷): share equally, split, each group, how many groups, per

Find more at
ViewMath.com/NC-Grade3

⊞ Multiplication Table ⊞

×	1	2	3	4	5	6	7	8	9	10	11
1	1	2	3	4	5	6	7	8	9	10	11
2	2	4	6	8	10	12	14	16	18	20	22
3	3	6	9	12	15	18	21	24	27	30	33
4	4	8	12	16	20	24	28	32	36	40	44
5	5	10	15	20	25	30	35	40	45	50	55
6	6	12	18	24	30	36	42	48	54	60	66
7	7	14	21	28	35	42	49	56	63	70	77
8	8	16	24	32	40	48	56	64	72	80	88
9	9	18	27	36	45	54	63	72	81	90	99
10	10	20	30	40	50	60	70	80	90	100	110
11	11	22	33	44	55	66	77	88	99	110	121

💡 How to Use This Table

To find 4×7:

1. Find **4** in the left column (blue).
2. Find **7** in the top row (blue).
3. Follow the row and column until they meet: the answer is **28**!

📈 My Confidence Tracker 📈

Record your scores below. You'll be amazed at your progress!

My name: ______________________________________

📋 Test	📅 Date	⭐ Score	🙂 How I Feel
1			
2			
3			

The easiest topic for me was:

____

The trickiest topic for me was:

____

One thing I got better at from Test 1 to Test 3:

____

Next time I want to try:

____

 *You just finished 3 practice tests — that's awesome! Compare your first score to your last. I bet you'll see real improvement. Ready for more? Check out our 5-test or 7-test books for even more practice!*

Find more at
ViewMath.com/NC-Grade3

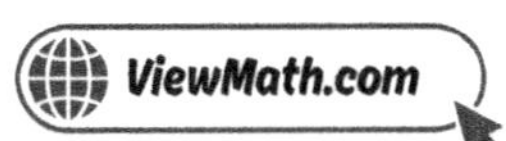

Continue Learning at ViewMath Academy!

For Parents, Teachers & Students

Great job on the practice tests! Want to keep improving? ViewMath Academy is your **free online companion** to this book.

- **Score Analyzer** — Enter your answers and instantly see which topics need more practice

- **Interactive Lessons** — Review the concepts behind each question with clear explanations

- **Adaptive Quizzes** — Practice your weak topics with questions that match your level

- **Progress Tracking** — See your mastery grow across all Grade 3 math topics

- **Personalized Dashboard** — A learning plan tailored just for you

Scan to visit ViewMath Academy

viewmath.com/academy

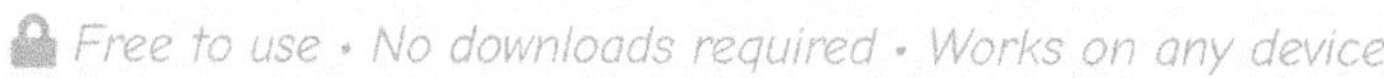 Free to use · No downloads required · Works on any device 

 Table of Contents

Here's what we'll explore together!

 *Let's learn and have fun!*

Practice Test 1

📋 30 Questions

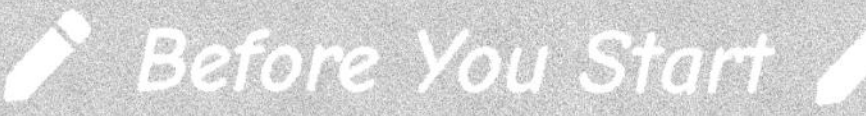

✏️ Before You Start ✏️

- ✔ **Read each question carefully** before choosing your answer.
- ✔ **Show your work** on scratch paper when you need to.
- ✔ **Skip hard questions** and come back to them later.
- ✔ **Check your answers** when you're done.
- ✔ **Take your time** — there's no rush!

⭐ You've Got This! ⭐

Do your best and show what you know!

1. Which number has a 4 in the hundreds place and a 7 in the thousands place?

   (A) 4,731

   (B) 7,431

   (C) 7,341

   (D) 3,470

2. Which number rounds to 500 when rounded to the nearest 100?

   (A) 428

   (B) 449

   (C) 462

   (D) 551

3. Which group contains only even numbers?

   (A) 12, 34, 57

   (B) 20, 46, 88

   (C) 31, 50, 72

   (D) 14, 63, 90

4. Marcus says $276 + 358 = 534$. What mistake did he most likely make?

   (A) He forgot to regroup in the ones column

   (B) He forgot to regroup in the tens column

   (C) He added the hundreds digits incorrectly

   (D) He subtracted instead of adding

5. What is $6,500 + 4,500$?

   (A) 10,000

   (B) 10,100

   (C) 11,000

   (D) 11,100

6. Estimate $248 + 175 + 362$ by rounding each number to the nearest hundred.

   Your Answer

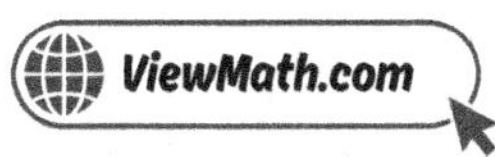

7. A spider has 8 legs. How many legs do 6 spiders have altogether? Write a multiplication sentence and solve.

Your Answer:

8. Mia says $3 \times 9 = 29$. What is the correct answer?

(A) 26

(B) 27

(C) 28

(D) 30

9. What is $8 \div 8$?

Your Answer:

10. Which set of numbers does NOT form a fact family?

(A) $3, 9, 27$

(B) $6, 7, 42$

(C) $4, 5, 25$

(D) $8, 8, 64$

11. Odd $\times$ odd $= ?$

(A) Always even

(B) Always odd

(C) Sometimes even, sometimes odd

(D) Always 0

12. Which fraction has a numerator of 2 and a denominator of 6?

(A) $\frac{6}{2}$

(B) $\frac{2}{6}$

(C) $\frac{2}{2}$

(D) $\frac{6}{6}$

Find more at
ViewMath.com/NC-Grade3

13. Is $\frac{3}{3}$ at 0 or at 1 on the number line?

Your Answer:

14. What is special about ALL unit fractions?

(A) They all have the same denominator

(B) They all equal 1

(C) They all have 1 as the numerator

(D) They are all less than $\frac{1}{2}$

15. $\frac{2}{4} = \frac{?}{2}$. What is the missing numerator?

Your Answer:

16. Write 4 as a fraction with denominator 3.

(A) $\frac{4}{3}$

(B) $\frac{3}{4}$

(C) $\frac{7}{3}$

(D) $\frac{12}{3}$

17. Tom ate $\frac{3}{8}$ of a pizza. Mia ate $\frac{5}{8}$ of the same pizza. Who ate more?

(A) Tom

(B) Mia

(C) They ate the same amount

(D) Cannot tell

18. You eat dinner at 6:30 in the evening. Is this A.M. or P.M.?

Your Answer

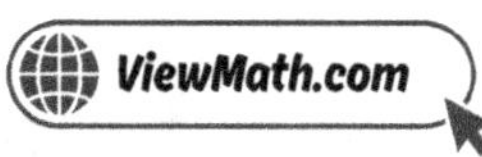

19. Emma started reading at 3:30 and stopped at 4:10. How long did she read?

    (A) 30 minutes

    (B) 40 minutes

    (C) 50 minutes

    (D) 1 hour and 10 minutes

20. 9,000 grams = how many kilograms?

    (A) 9 kg

    (B) 90 kg

    (C) 900 kg

    (D) 9,000 kg

21. A pitcher holds 8 liters of juice. The family drinks 3 liters. How much juice is left?

    (A) 3 L

    (B) 5 L

    (C) 8 L

    (D) 11 L

22. You have 3 quarters, 2 dimes, and 1 nickel. How much money do you have in cents?

    Your Answer

23. What is $2.50 + $1.25?

    (A) $3.25

    (B) $3.75

    (C) $4.75

    (D) $3.55

24. A picture graph has the key: Each ★ = 5 books. The row for Jake has 3 stars. How many books did Jake read?

    (A) 3

    (B) 5

    (C) 8

    (D) 15

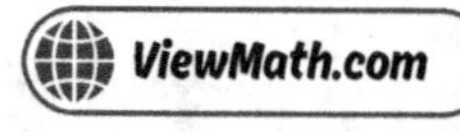

25. A line plot shows pencil lengths. There are 3 X marks above $1\frac{1}{2}$ inches. What does this mean?

(A) 3 pencils are $1\frac{1}{2}$ inches long

(B) $1\frac{1}{2}$ pencils are 3 inches long

(C) The total length is $4\frac{1}{2}$ inches

(D) There are 3 pencils in all

26. Which statement about a rectangle is TRUE?

(A) All 4 sides are always equal.

(B) It has exactly 3 right angles.

(C) It has 4 right angles and 2 pairs of equal sides.

(D) It has no parallel sides.

27. How many flat faces does a cone have?

Your Answer:

28. A square has sides that are 7 inches long. What is its area?

(A) 14 sq in

(B) 28 sq in

(C) 42 sq in

(D) 49 sq in

29. Lily says a rectangle that is 5 cm long and 3 cm wide has a perimeter of 15 cm. What mistake did she make?

(A) She subtracted instead of adding.

(B) She found the area instead of the perimeter.

(C) She forgot to add all 4 sides.

(D) She divided instead of multiplying.

30. If one pizza is cut into 4 equal slices and another identical pizza is cut into 8 equal slices, which slices are bigger?

(A) The $\frac{1}{8}$ slices

(B) The $\frac{1}{4}$ slices

(C) They are the same size.

(D) You cannot tell.

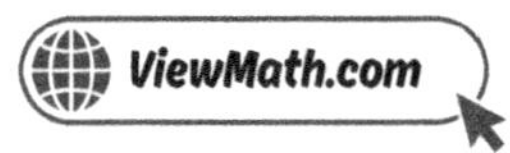

 # End of Practice Test 1 

Great job finishing the test!

 My Score

I got ___________ out of 30 questions right.

*Check your answers in the **Answer Key** at the back of the book.*

Review any questions you missed. That's how we learn!

Check Your Score Online!

Visit **ViewMath Academy** to enter your answers and see which topics you need to review. You can also explore lessons, take quizzes, track your scores, and save your progress!

viewmath.com/score/3.1.NC.01

Or go to viewmath.com/score and enter code: 3.1.NC.01

2

Practice Test 2

30 Questions

✏️ Before You Start ✏️

✔ **Read each question carefully** before choosing your answer.

✔ **Show your work** on scratch paper when you need to.

✔ **Skip hard questions** and come back to them later.

✔ **Check your answers** when you're done.

✔ **Take your time** — there's no rush!

⭐ You've Got This! ⭐

Do your best and show what you know!

1. A number has 3 thousands, 7 hundreds, 0 tens, and 5 ones. What is the number?

Your Answer:

2. Maria rounded 438 to the nearest 10 and got 430. Is she correct?

(A) No, it should be 440

(B) Yes, 438 rounds to 430

(C) No, it should be 400

(D) No, it should be 450

3. List all the even numbers between 21 and 30.

Your Answer:

4. What is $198 + 345 + 267$?

Your Answer:

5. What is $4,008 + 3,995$?

(A) 7,003

(B) 7,903

(C) 7,993

(D) 8,003

6. Estimate $358 + 246$ by rounding each number to the nearest hundred.

Your Answer:

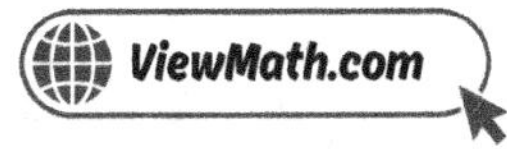

7. Which skip counting sequence helps you find 4×3?

(A) $3, 6, 9, 12$

(B) $4, 8, 12$

(C) $3, 4, 5, 6$

(D) $1, 2, 3, 4$

8. There are 3 tricycles in a park. Each tricycle has 3 wheels. How many wheels are there in all?

(A) 6

(B) 9

(C) 12

(D) 15

9. What is $20 \div 4$?

(A) 4

(B) 5

(C) 16

(D) 24

10. Write the missing fact in this fact family: $7 \times 6 = 42$, $6 \times 7 = 42$, $42 \div 7 = 6$, ___

(A) $42 \div 6 = 7$

(B) $42 \div 42 = 1$

(C) $7 + 6 = 13$

(D) $42 - 7 = 35$

11. Looking at the diagonal of a multiplication table ($1 \times 1, 2 \times 2, 3 \times 3, \ldots$), what are these numbers called?

(A) Even numbers

(B) Odd numbers

(C) Square numbers

(D) Prime numbers

12. A pizza is cut into 8 equal slices. You eat 3 slices. What fraction of the pizza did you eat?

(A) $\frac{3}{5}$

(B) $\frac{3}{8}$

(C) $\frac{8}{3}$

(D) $\frac{5}{8}$

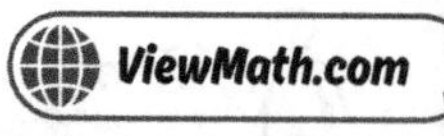

13. On a number line divided into 8 parts, what fraction names the fifth mark from 0?

(A) $\frac{5}{8}$

(B) $\frac{8}{5}$

(C) $\frac{3}{8}$

(D) $\frac{5}{5}$

14. Which list shows unit fractions from smallest to largest?

(A) $\frac{1}{2}, \frac{1}{4}, \frac{1}{8}$

(B) $\frac{1}{8}, \frac{1}{4}, \frac{1}{2}$

(C) $\frac{1}{4}, \frac{1}{2}, \frac{1}{8}$

(D) $\frac{1}{8}, \frac{1}{2}, \frac{1}{4}$

15. Which pair of fractions are equivalent?

(A) $\frac{1}{4}$ and $\frac{2}{8}$

(B) $\frac{1}{4}$ and $\frac{3}{8}$

(C) $\frac{2}{3}$ and $\frac{3}{4}$

(D) $\frac{1}{3}$ and $\frac{1}{4}$

16. Which of these is a way to write 5 as a fraction?

(A) $\frac{5}{1}$

(B) $\frac{10}{2}$

(C) $\frac{15}{3}$

(D) All of the above

17. Is $\frac{5}{8}$ more or less than $\frac{1}{2}$?

(A) Less than $\frac{1}{2}$

(B) Equal to $\frac{1}{2}$

(C) More than $\frac{1}{2}$

(D) Cannot tell

18. You go to soccer practice at 4:00 in the afternoon. Is this A.M. or P.M.?

Your Answer:

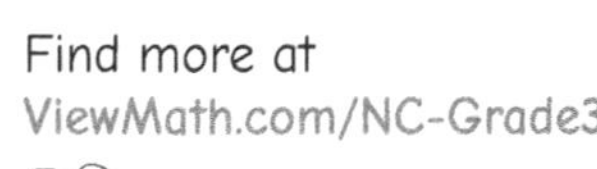

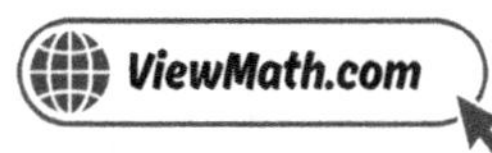

19. Jake starts his homework at 3:45 P.M. He works for 1 hour and 20 minutes. What time does he finish?

Your Answer:

20. A bag of rice has a mass of 5 kg. You use 2 kg for cooking. How much rice is left?

Your Answer:

21. Which container holds the LEAST liquid?

(A) A swimming pool

(B) A bathtub

(C) A bucket

(D) A teaspoon

22. Which group of coins equals 50 cents?

(A) 5 pennies

(B) 5 dimes

(C) 3 dimes

(D) 3 quarters

23. Lily has $6.00. She buys a sticker pack for $2.35 and a pen for $1.80. Does she have enough money left to buy a snack for $2.00?

(A) Yes, she has $2.85 left

(B) Yes, she has $2.00 left

(C) No, she only has $1.85 left

(D) No, she only has $1.65 left

24. A picture graph tracks books read in a month. Each symbol stands for 2 books. Amy has 5 symbols, Ben has 8 symbols, and Chloe has 3 symbols. How many books did they read altogether?

Your Answer:

Find more at
ViewMath.com/NC-Grade3

25. A line plot shows the lengths of 10 crayons. The data is shown below:

2 in.	$2\frac{1}{2}$ in.	3 in.	$3\frac{1}{2}$ in.	4 in.
2	3	4	1	0

How many crayons are shorter than 3 inches?

(A) 2

(B) 4

(C) 5

(D) 9

26. Every square is also a special kind of which shape?

(A) Triangle

(B) Pentagon

(C) Rectangle

(D) Trapezoid

27. How many edges does a rectangular prism have?

Your Answer:

28. A square has sides of 9 inches. What is the area?

Your Answer:

29. A triangle has sides of 4 cm, 6 cm, and 8 cm. What is the perimeter?

(A) 14 cm

(B) 18 cm

(C) 20 cm

(D) 24 cm

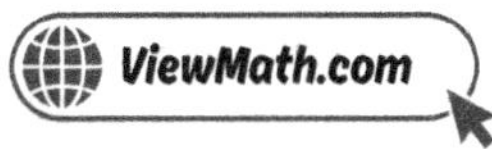

30. *You fold a piece of paper in half, then fold it in half again. How many equal parts do you have?*

Your Answer:

 End of Practice Test 2 

Great job finishing the test!

✅ My Score

I got _____________ out of 30 questions right.

Check your answers in the **Answer Key** at the back of the book.

💡 Review any questions you missed. That's how we learn!

📊 Check Your Score Online!

Visit **ViewMath Academy** to enter your answers and see which topics you need to review. You can also explore lessons, take quizzes, track your scores, and save your progress!

viewmath.com/score/3.1.NC.02

Or go to viewmath.com/score and enter code: 3.1.NC.02

3

Practice Test 3

 30 Questions

✏️ Before You Start ✏️

- ✓ **Read each question carefully** before choosing your answer.
- ✓ **Show your work** on scratch paper when you need to.
- ✓ **Skip hard questions** and come back to them later.
- ✓ **Check your answers** when you're done.
- ✓ **Take your time** — there's no rush!

⭐ You've Got This! ⭐

Do your best and show what you know!

1. Which number is the same as 9 thousands, 0 hundreds, 2 tens, and 5 ones?

(A) 9,250

(B) 9,025

(C) 9,205

(D) 9,520

2. When you round 549 to the nearest 10 and to the nearest 100, which gives the larger answer?

(A) Rounded to the nearest 10

(B) Rounded to the nearest 100

(C) They are the same

(D) Cannot tell without rounding

3. If you add two odd numbers, the result is always:

(A) Odd

(B) Even

(C) Greater than 10

(D) An odd number less than 20

4. A store sold 256 books on Monday and 478 books on Tuesday. How many books were sold in all?

(A) 624

(B) 634

(C) 724

(D) 734

5. A plane flew 1,892 miles on Saturday and 2,347 miles on Sunday. How far did it fly in total?

Your Answer

6. Which rounding method gives a closer estimate for $463 + 219$?

(A) Rounding to the nearest 10

(B) Rounding to the nearest 100

(C) Both give the same estimate

(D) You cannot estimate this sum

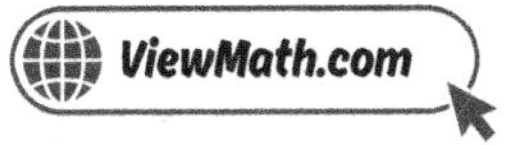

7. *How many stars are in 5 groups of 6?*

(A) 11

(B) 25

(C) 30

(D) 56

8. *What is 2×9?*

(A) 11

(B) 16

(C) 18

(D) 29

9. *What is $9 \div 1$?*

(A) 0

(B) 1

(C) 9

(D) 10

10. *Which multiplication fact helps solve $56 \div 8$?*

(A) $8 \times 6 = 48$

(B) $8 \times 7 = 56$

(C) $8 \times 8 = 64$

(D) $8 \times 9 = 72$

11. *The rule for a pattern is "multiply by 3." If the first number is 2, what are the next two numbers?*

(A) $5, 8$

(B) $4, 8$

(C) $6, 12$

(D) $6, 18$

12. *In the fraction $\frac{4}{6}$, what is the numerator?*

Your Answer:

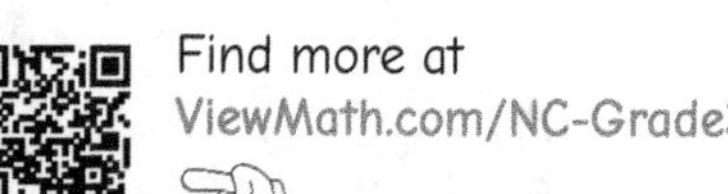
Find more at
ViewMath.com/NC-Grade3
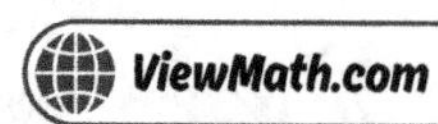

13. A number line from 0 to 1 is divided into 6 equal parts. Which fraction is closest to 1?

(A) $\frac{1}{6}$

(B) $\frac{3}{6}$

(C) $\frac{4}{6}$

(D) $\frac{5}{6}$

14. Which list shows unit fractions in order from largest to smallest?

(A) $\frac{1}{8}, \frac{1}{6}, \frac{1}{4}, \frac{1}{3}$

(B) $\frac{1}{3}, \frac{1}{4}, \frac{1}{6}, \frac{1}{8}$

(C) $\frac{1}{4}, \frac{1}{8}, \frac{1}{3}, \frac{1}{6}$

(D) $\frac{1}{2}, \frac{1}{8}, \frac{1}{4}, \frac{1}{6}$

15. Write two fractions that are equivalent to $\frac{1}{4}$.

Your Answer:

16. On a number line divided into fourths, which fraction is at 3?

(A) $\frac{3}{4}$

(B) $\frac{4}{3}$

(C) $\frac{12}{4}$

(D) $\frac{3}{1}$

17. When two fractions have the same denominator, how do you compare them?

(A) The one with the bigger denominator is larger

(B) The one with the bigger numerator is larger

(C) They are always equal

(D) You cannot compare them

18. A clock shows the short hand past the 10 and the long hand between the 2 and 3, on the second tick mark past the 2. What time does the clock show?

(A) 10:10

(B) 10:12

(C) 2:50

(D) 10:15

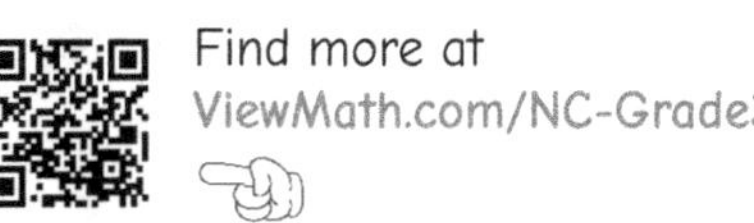

Find more at
ViewMath.com/NC-Grade3

19. A library is open from 10:00 A.M. to 3:00 P.M. How long is it open?

(A) 3 hours

(B) 4 hours

(C) 5 hours

(D) 7 hours

20. A box of cereal has a mass of 500 g. You buy 2 boxes. What is the total mass?

(A) 502 g

(B) 700 g

(C) 1,000 g

(D) 5,000 g

21. A pot holds 6 liters when full. You pour in 2 liters first, then 3 more liters. How many more liters are needed to fill the pot?

Your Answer:

22. How much is a quarter worth?

(A) 1 cent

(B) 5 cents

(C) 10 cents

(D) 25 cents

23. What is $3.45 + $2.80?

(A) $5.25

(B) $6.25

(C) $5.65

(D) $6.15

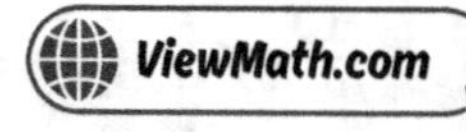

24. *A picture graph shows toys collected. Each symbol stands for 5 toys.*

Bears: ★★★
Cars: ★★★★★
Dolls: ★★

How many toys were collected in all?

(A) 10

(B) 25

(C) 50

(D) 75

25. *A line plot shows the lengths of nails in inches. The data points are: 1, 1, $1\frac{1}{2}$, 2, 2, 2, $2\frac{1}{2}$, $2\frac{1}{2}$. How many X marks should be placed above 2 inches?*

26. *Mia says, "All rectangles are squares." Is she correct?*

(A) *Yes, because both have 4 sides.*

(B) *Yes, because both have 4 right angles.*

(C) *No, because a rectangle does not always have 4 equal sides.*

(D) *No, because a rectangle has no right angles.*

27. *Two cubes are placed side by side. How many faces do they have in total?*

(A) 6

(B) 8

(C) 10

(D) 12

28. *A rectangle is 10 feet long and 3 feet wide. What is its area?*

(A) *13 sq ft*

(B) *26 sq ft*

(C) *30 sq ft*

(D) *33 sq ft*

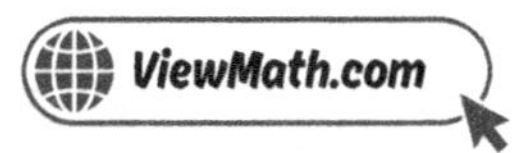

29. Which shape has the greatest perimeter?

(A) A square with sides of 5 cm

(B) A rectangle that is 8 cm × 3 cm

(C) A rectangle that is 6 cm × 4 cm

(D) A triangle with sides 7 cm, 7 cm, and 7 cm

30. A square is partitioned into 4 equal parts. All 4 parts are shaded. What fraction is shaded?

(A) $\frac{1}{4}$

(B) $\frac{3}{4}$

(C) $\frac{4}{4}$

(D) $\frac{4}{1}$

 End of Practice Test 3 

Great job finishing the test!

 My Score

I got _____________ out of 30 questions right.

Check your answers in the **Answer Key** at the back of the book.

💡 *Review any questions you missed. That's how we learn!*

📊 Check Your Score Online!

Visit **ViewMath Academy** to enter your answers and see which topics you need to review. You can also explore lessons, take quizzes, track your scores, and save your progress!

viewmath.com/score/3.1.NC.03

Or go to viewmath.com/score and enter code: 3.1.NC.03

Answer Key & Explanations

⭐ Check Your Answers! ⭐

First try each test on your own, then look here to check.

Read the explanations to learn from any mistakes ⭐

📋 Practice Test 1 — Answer Key

1 B	**2** C	**3** B	**4** B	**5** C	**6** 800	**7** $6 \times 8 = 48$	**8** B	**9** 1
10 C	**11** B	**12** B	**13** 1	**14** C	**15** 1	**16** D	**17** B	**18** P.M.	**19** B
20 A	**21** B	**22** 100 *cents* (or $1.00)	**23** B	**24** D	**25** A	**26** C	**27** 1
28 D	**29** B	**30** B						

💡 Time to Learn! 💡

*Go through the explanations below, **especially for the questions you missed**.*

Understanding why each answer is correct makes you a stronger math thinker!

👍 **Tip:** *Circle any questions you got wrong, then read their explanation carefully.*

📖 Practice Test 1 — Detailed Explanations

1. *In 7,431: the 7 is in the thousands place and the 4 is in the hundreds place.*

2   462 has tens digit $6 \geq 5$, so it rounds up to 500. 428 and 449 round to 400. 551 rounds to 600.

3   20 (ends in 0), 46 (ends in 6), 88 (ends in 8) are all even. The other groups each contain at least one odd number.

4   The correct answer is $276 + 358 = 634$. Ones: $6 + 8 = 14$, carry 1. Tens: $7 + 5 + 1 = 13$, carry 1. Hundreds: $2 + 3 + 1 = 6$. Marcus got 534, meaning his hundreds digit is 5 instead of 6. He likely forgot to add the carry from the tens column to the hundreds.

5   $6{,}500 + 4{,}500 = 11{,}000$. Two 4-digit numbers can sometimes add up to a 5-digit number!

6   $248 \approx 200$, $175 \approx 200$, and $362 \approx 400$. So $200 + 200 + 400 = 800$.

7   6 spiders with 8 legs each gives $6 \times 8 = 48$ legs.

8   $3 \times 9 = 9 + 9 + 9 = 27$. Mia's mistake was close, but the correct product is 27.

9   Any number divided by itself equals 1.

10   $4 \times 5 = 20$, not 25. So $4, 5, 25$ do not form a fact family.

11   Odd $\times$ odd always gives an odd product. Example: $3 \times 5 = 15$, $7 \times 9 = 63$.

12   Numerator is on top and denominator is on the bottom. So it is $\frac{2}{6}$.

13   $\frac{3}{3}$ means all 3 parts, which equals 1 whole.

14   A unit fraction always has 1 as the numerator. Examples: $\frac{1}{2}, \frac{1}{3}, \frac{1}{4}, \frac{1}{8}$.

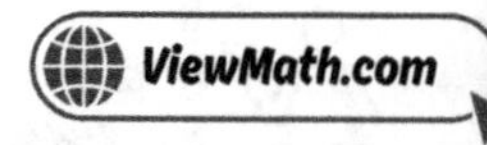

15. Divide top and bottom by 2: $\frac{2 \div 2}{4 \div 2} = \frac{1}{2}$.

16. $4 \times 3 = 12$ thirds. So $4 = \frac{12}{3}$.

17. Same denominator. $5 > 3$, so $\frac{5}{8} > \frac{3}{8}$. Mia ate more.

18. Evening is after noon, so 6:30 in the evening is 6:30 P.M.

19. From 3:30 to 4:00 is 30 minutes. From 4:00 to 4:10 is 10 minutes. Total: $30 + 10 = 40$ minutes.

20. Divide by 1,000: $9,000 \div 1,000 = 9$ kg.

21. $8 - 3 = 5$ L.

22. $3 \times 25 = 75$. Then $2 \times 10 = 20$. Then $1 \times 5 = 5$. Total: $75 + 20 + 5 = 100$ cents.

23. Line up the decimals and add: $\$2.50 + \$1.25 = \$3.75$.

24. Each star $= 5$ books. Jake has 3 stars, so $3 \times 5 = 15$ books.

25. Each X mark represents one pencil. 3 X marks above $1\frac{1}{2}$ means 3 pencils measured $1\frac{1}{2}$ inches.

26. A rectangle has 4 right angles and 2 pairs of equal sides (opposite sides are equal). Not all 4 sides have to be equal.

27. A cone has 1 flat circular face on the bottom and 1 curved surface that comes to a point at the top.

28. A square is a rectangle with all sides equal. Area $= 7 \times 7 = 49$ sq in.

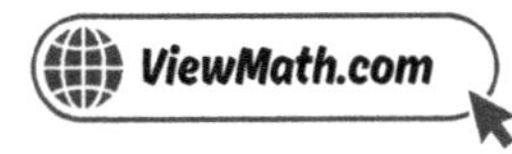

(29) Lily multiplied $5 \times 3 = 15$, which gives the area, not the perimeter. The correct perimeter is $5+3+5+3 = 16$ cm.

(30) The $\frac{1}{4}$ slices are bigger because the pizza is cut into fewer pieces. More equal parts means smaller pieces: $\frac{1}{4} > \frac{1}{8}$.

✅ Practice Test 2 — Answer Key

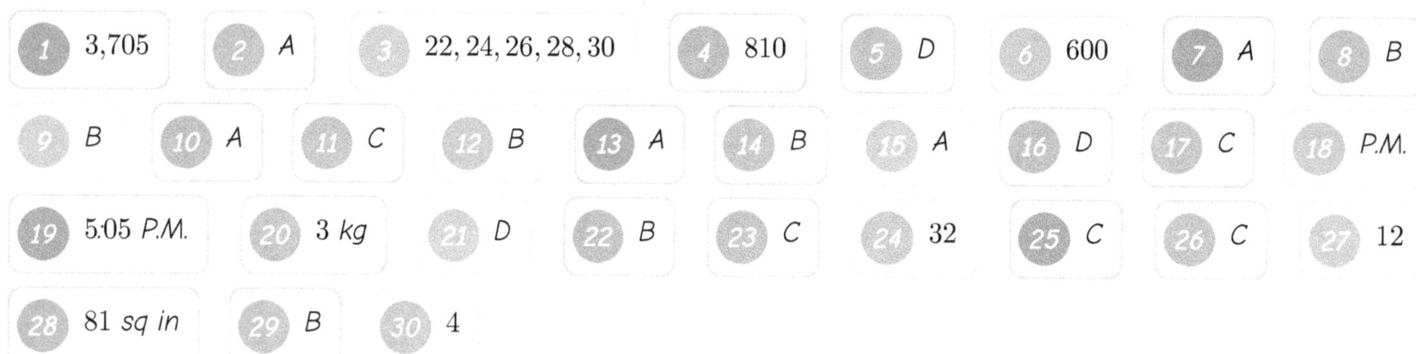

1	3,705	2	A	3	22, 24, 26, 28, 30	4	810	5	D	6	600	7	A	8	B
9	B	10	A	11	C	12	B	13	A	14	B	15	A	16	D	17	C	18	P.M.
19	5:05 P.M.	20	3 kg	21	D	22	B	23	C	24	32	25	C	26	C	27	12
28	81 sq in	29	B	30	4

💡 Time to Learn! 💡

Go through the explanations below, **especially for the questions you missed.**

Understanding why each answer is correct makes you a stronger math thinker!

👍 **Tip:** Circle any questions you got wrong, then read their explanation carefully.

📖 Practice Test 2 — Detailed Explanations

(1) 3 thousands $= 3,000$, 7 hundreds $= 700$, 0 tens $= 0$, 5 ones $= 5$. The number is $3,000 + 700 + 5 = 3,705$.

(2) The ones digit is 8. Since $8 \geq 5$, we round up. 438 rounds to 440, not 430.

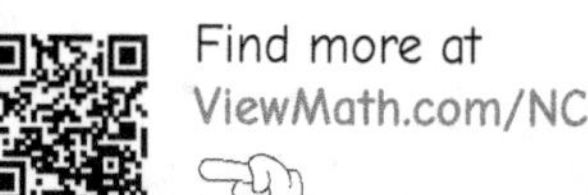

3. The even numbers between 21 and 30 are $22, 24, 26, 28, 30$. They all end in an even digit.

4. First add $198 + 345 = 543$. Ones: $8 + 5 = 13$, carry 1. Tens: $9 + 4 + 1 = 14$, carry 1. Hundreds: $1 + 3 + 1 = 5$. Then $543 + 267 = 810$. Ones: $3 + 7 = 10$, carry 1. Tens: $4 + 6 + 1 = 11$, carry 1. Hundreds: $5 + 2 + 1 = 8$.

5. Ones: $8 + 5 = 13$, carry 1. Tens: $0 + 9 + 1 = 10$, carry 1. Hundreds: $0 + 9 + 1 = 10$, carry 1. Thousands: $4 + 3 + 1 = 8$. The sum is $8,003$. Watch out for chains of carries through zeros!

6. $358 \approx 400$ and $246 \approx 200$. So $400 + 200 = 600$.

7. 4×3 means count by 3s four times: $3, 6, 9, 12$. So $4 \times 3 = 12$.

8. $3 \times 3 = 9$ wheels.

9. 20 split into 4 equal groups gives 5 in each group. $20 \div 4 = 5$.

10. The missing fact is $42 \div 6 = 7$. A fact family with different factors has 4 facts.

11. Numbers like $1, 4, 9, 16, 25, 36, \ldots$ where a number is multiplied by itself are called square numbers.

12. You ate 3 out of 8 equal slices, so the fraction is $\frac{3}{8}$.

13. Counting 5 marks from 0 on a line divided into 8 parts gives $\frac{5}{8}$.

14. Bigger denominators mean smaller pieces: $\frac{1}{8} < \frac{1}{4} < \frac{1}{2}$.

15. $\frac{1}{4} = \frac{1 \times 2}{4 \times 2} = \frac{2}{8}$.

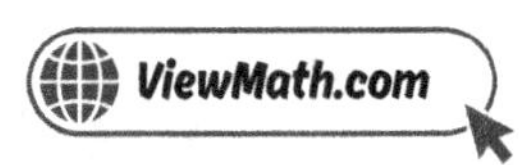

16 $\frac{5}{1} = 5$, $\frac{10}{2} = 5$, and $\frac{15}{3} = 5$. They all equal 5.

17 $\frac{1}{2} = \frac{4}{8}$. Since $\frac{5}{8} > \frac{4}{8}$, it is more than $\frac{1}{2}$.

18 Afternoon is after noon, so 4:00 in the afternoon is 4:00 P.M.

19 $3:45 + 1$ hour $= 4:45$. Then $4:45 + 20$ minutes $= 5:05$ P.M.

20 $5 - 2 = 3$ kg.

21 A teaspoon holds much less than 1 liter, which is far less than a bucket, bathtub, or pool.

22 5 dimes $= 5 \times 10 = 50$ cents.

23 Total spent: $\$2.35 + \$1.80 = \$4.15$. Money left: $\$6.00 - \$4.15 = \$1.85$. Since $\$1.85 < \$2.00$, she does not have enough.

24 Amy: $5 \times 2 = 10$. Ben: $8 \times 2 = 16$. Chloe: $3 \times 2 = 6$. Total: $10 + 16 + 6 = 32$ books.

25 Crayons shorter than 3 inches: 2 in. (2) $+ 2\frac{1}{2}$ in. (3) $= 5$ crayons.

26 A square has 4 right angles just like a rectangle, so every square is a special kind of rectangle. It also has 4 equal sides, making it a special rhombus too.

27 A rectangular prism has 12 edges, just like a cube. Edges are the lines where two faces meet.

28 Area $= 9 \times 9 = 81$ sq in.

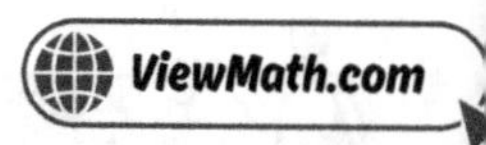

29. Add all side lengths: $4 + 6 + 8 = 18$ cm.

30. Folding in half gives 2 parts. Folding in half again doubles it to 4 equal parts. Each part is $\frac{1}{4}$ of the whole.

✔ Practice Test 3 — Answer Key

1 B	**2** A	**3** B	**4** D	**5** 4,239	**6** A	**7** C	**8** C	**9** C	**10** B
11 D	**12** 4	**13** D	**14** B	**15** $\frac{2}{8}$ and $\frac{3}{12}$	**16** C	**17** B	**18** B	**19** C	
20 C	**21** 1 L	**22** D	**23** B	**24** C	**25** 3	**26** C	**27** D	**28** C	**29** B
30 C									

💡 Time to Learn! 💡

Go through the explanations below, **especially for the questions you missed**.

Understanding why each answer is correct makes you a stronger math thinker!

👍 **Tip:** Circle any questions you got wrong, then read their explanation carefully.

📖 Practice Test 3 — Detailed Explanations

1. 9 thousands $= 9,000$, 0 hundreds $= 0$, 2 tens $= 20$, 5 ones $= 5$. So $9,000 + 20 + 5 = 9,025$.

2. 549 rounded to the nearest 10: ones digit is $9 \geq 5$, so 550. Rounded to the nearest 100: tens digit is $4 < 5$, so 500. $550 > 500$.

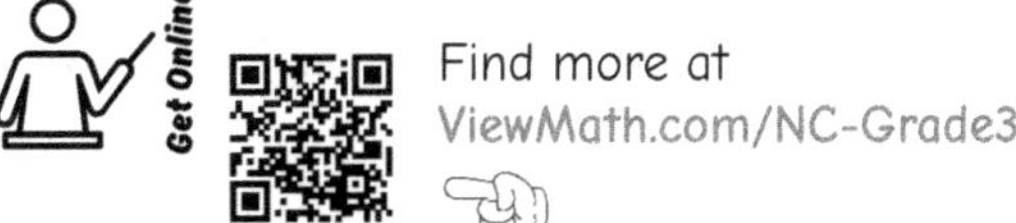

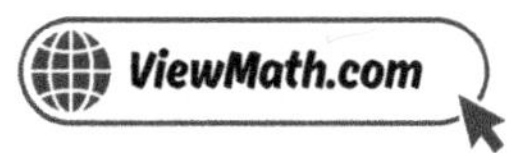

3 $Odd + Odd = Even$. For example, $3 + 5 = 8$ (even) and $7 + 9 = 16$ (even).

4 $256 + 478 = 734$. Ones: $6 + 8 = 14$, carry 1. Tens: $5 + 7 + 1 = 13$, carry 1. Hundreds: $2 + 4 + 1 = 7$.

5 $1{,}892 + 2{,}347 = 4{,}239$. Ones: $2 + 7 = 9$. Tens: $9 + 4 = 13$, carry 1. Hundreds: $8 + 3 + 1 = 12$, carry 1. Thousands: $1 + 2 + 1 = 4$.

6 Nearest 10: $460 + 220 = 680$. Nearest 100: $500 + 200 = 700$. The exact answer is 682, so rounding to the nearest 10 gives a closer estimate.

7 5 groups of 6 means $5 \times 6 = 30$ stars.

8 $2 \times 9 = 9 + 9 = 18$. Multiplying by 2 is doubling.

9 Any number divided by 1 equals itself. $9 \div 1 = 9$.

10 $8 \times 7 = 56$, so $56 \div 8 = 7$.

11 $2 \times 3 = 6$, then $6 \times 3 = 18$. The next two numbers are $6, 18$.

12 The numerator is the top number. In $\frac{4}{6}$, the numerator is 4.

13 $\frac{5}{6}$ is the last mark before 1, so it is closest to 1.

14 Largest to smallest: bigger pieces first. $\frac{1}{3} > \frac{1}{4} > \frac{1}{6} > \frac{1}{8}$.

15 $\frac{1 \times 2}{4 \times 2} = \frac{2}{8}$ and $\frac{1 \times 3}{4 \times 3} = \frac{3}{12}$.

Find more at
ViewMath.com/NC-Grade3

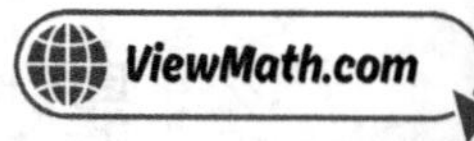

16   $3 = 3 \times \frac{4}{4} = \frac{12}{4}$. On a fourths number line, 3 is at $\frac{12}{4}$.

17   Same denominator means same-size pieces. More pieces (bigger numerator) = bigger fraction.

18   The short hand past 10 means the hour is 10. The long hand on 2 means 10 minutes. Two more tick marks gives $10 + 2 = 12$ minutes. The time is 10:12.

19   From 10:00 A.M. to 3:00 P.M.: count $10 \to 11 \to 12 \to 1 \to 2 \to 3$, which is 5 hours.

20   $500 + 500 = 1{,}000$ g (which is the same as 1 kg).

21   You poured in $2 + 3 = 5$ L. The pot holds 6 L, so $6 - 5 = 1$ more liter is needed.

22   A quarter is worth 25 cents.

23   Cents: $45 + 80 = 125$ cents $= 1$ dollar and 25 cents. Dollars: $3 + 2 + 1 = 6$. Answer: \$6.25.

24   Bears: $3 \times 5 = 15$. Cars: $5 \times 5 = 25$. Dolls: $2 \times 5 = 10$. Total: $15 + 25 + 10 = 50$ toys.

25   Count how many times 2 appears in the data: 2, 2, 2. That's 3 times.

26   A square must have 4 equal sides AND 4 right angles. A rectangle has 4 right angles but does not always have 4 equal sides (only when it is a square).

27   Each cube has 6 faces. Two cubes have $6 + 6 = 12$ faces in total.

28   Area $= 10 \times 3 = 30$ sq ft.

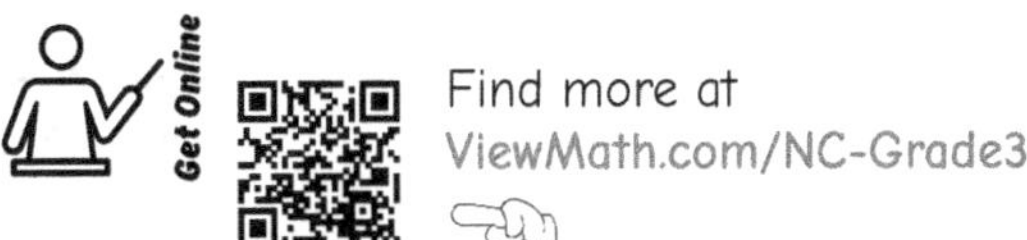
Find more at
ViewMath.com/NC-Grade3

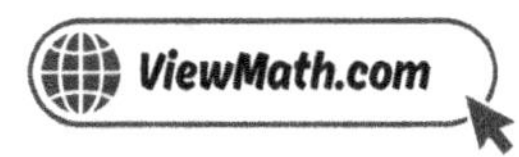

29 Square: $4 \times 5 = 20$ cm. Rectangle 8×3: $(2 \times 8) + (2 \times 3) = 22$ cm. Rectangle 6×4: $(2 \times 6) + (2 \times 4) = 20$ cm. Triangle: $7 + 7 + 7 = 21$ cm. The 8×3 rectangle has the greatest perimeter of 22 cm.

30 All 4 out of 4 parts are shaded, so $\frac{4}{4}$ is shaded. $\frac{4}{4} = 1$ whole.

Great job checking your work!

Keep practicing and you'll be a math star!